W9-BUP-979

Biosphere 2

Solving Word Problems

JESSUP ELEMENTARY

510 SHE

Shea, Therese
Biosphere 2 : solving word problems

3112271195

Therese Shea

PowerMath™

The Rosen Publishing Group's
PowerKids Press™
New York

Published in 2005 by The Rosen Publishing Group, Inc.
29 East 21st Street, New York, NY 10010

Copyright © 2005 by The Rosen Publishing Group, Inc.

All rights reserved. No part of this book may be reproduced in any form without permission
in writing from the publisher, except by a reviewer.

Book Design: Haley Wilson

Photo Credits: Cover, pp. 7, 9, 10, 13, 14, 17, 19, 21, 22, 25, 28 © Roger Ressmeyer/Corbis; p. 5 ©
Joseph Sohm; ChromoSohm Inc./Corbis; p. 26 © Craig Tuttle/Corbis; pp. 27, 30 © PhotoDisc.

Library of Congress Cataloging-in-Publication Data

Shea, Therese.
  Biosphere 2 : solving word problems / Therese Shea.
      p. cm. — (PowerMath)
  Includes index.
  ISBN 1-4042-2943-4 (lib. bdg.)
  ISBN 1-4042-5150-2 (pbk.)
  6-pack ISBN 1-4042-5151-0
  1. Problem solving—Juvenile literature. 2. Word problems (Mathematics)—Juvenile literature. 3.
Biosphere 2 (Project)—Juvenile literature. I. Title. II. Series.
  QA63.S44 2005
  510—dc22
                              2004005882

Manufactured in the United States of America

Contents

What Is Biosphere 2?

Have you ever heard of Biosphere 2? You might be thinking, "What was the first biosphere?" We should all know Biosphere 1 very well— Biosphere 1 is Earth! "Biosphere" is the name for living **organisms** and their **environment**. Earth is our largest example of plants, animals, and human beings living and growing together successfully.

A group of people who wanted to learn more about how organisms living together affect one another created Biosphere 2 as a smaller, contained version of Earth. They designed Biosphere 2 to keep life going inside the enclosed space without any outside assistance (except for sunlight and electricity). The Biosphere 2 researchers believed that the experiments carried out within its walls could help us understand how all the different elements in any biosphere—including Earth—affect each other.

The people responsible for creating and operating Biosphere 2 needed to solve many problems as they worked. Throughout this book, we will find answers to word problems like the ones these scientists might have solved to help us understand the results of the Biosphere 2 experiment. A word problem is a math question expressed in words. We encounter word problems every time we talk about or read about a situation that requires math to find an answer. When working on a word problem, we must first carefully read it to make sure we fully understand what the problem is asking. Then we need to write it as a mathematical equation to solve. Once we find the answer to the equation, we can express it using words. Read on to learn more about Biosphere 2 and word problems.

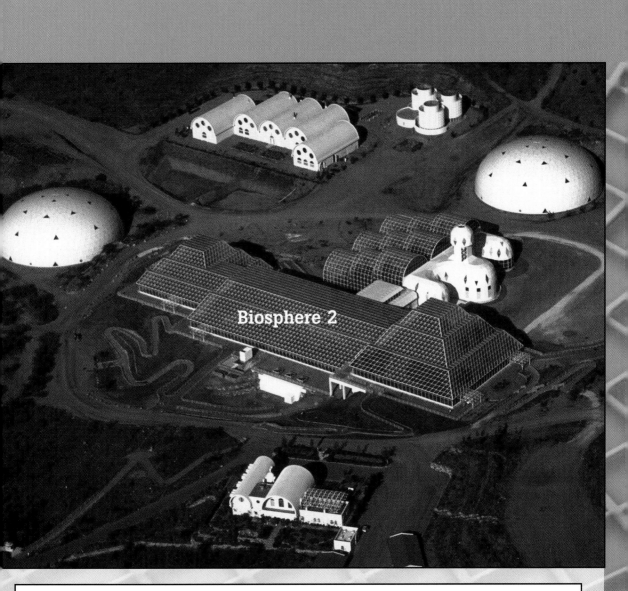

Biosphere 2

The Biosphere 2 scientists believed that if the experiment were successful, Biosphere 2 could be constructed again as a way for humans to live comfortably on the moon or even on Mars!

After the initial research was completed, the Biosphere 2 team needed money to start the project. Let's say that the group decided they needed $9 million for salaries, $75 million for construction materials, and $66 million to transport thousands of plants and animals to Biosphere 2. How much money did they need altogether? The word "altogether" tells us that we need to add these numbers together to find the answer.

$$
\begin{array}{r}
\$\ 9,000,000 \\
75,000,000 \\
+\ 66,000,000 \\
\hline
\$150,000,000
\end{array}
$$

The Biosphere 2 team needed $150 million to start the Biosphere 2 project.

In the late 1980s, a Texas oil billionaire named Edward Bass gave the Biosphere 2 researchers $150 million to start the project. The designers knew the structure had to be large enough to hold many different animals and ecological areas. Biosphere 2 was constructed of glass and steel in the Sonoran Desert about 30 miles from Tucson, Arizona.

Biosphere 2 covers 3.15 acres of desert. If 1 acre of land equals 43,560 square feet, how many square feet of desert does Biosphere 2 cover? We know that the structure covers 3.15 acres and that 1 acre of land equals 43,560 square feet. Therefore, we need to multiply the number of square feet in 1 acre by the number of acres to find the total number of square feet.

$$
\begin{array}{r}
43,560 \text{ square feet per acre} \\
\times\ \ \ 3.15 \text{ acres} \\
\hline
2\ 178\ 00 \\
4\ 356\ 0 \\
+\ 130,680 \\
\hline
137,214.00 \text{ square feet}
\end{array}
$$

Biosphere 2 covers 137,214 square feet.

The beams that hold Biosphere 2's glass panes together are called space frames. They are made of steel, but they are hollow and light and fit together easily to form the building's unusual shape.

Biosphere 2's glass walls made the building like a greenhouse and created problems. The inside air heated quickly under the desert sun. It expanded in volume, pushing the walls outward. Cool temperatures at night had the opposite effect. These problems were solved by building 2 giant "lungs" about 150 feet from Biosphere 2. During the day, heated air traveled from inside Biosphere 2 through underground tunnels to the lungs, which were made of a material that allowed them to expand. As the night air cooled and shrank in volume, the lungs contracted and sent cooler air back into Biosphere 2.

Inside Biosphere 2, scientists re-created 5 Earth **biomes**: ocean, marsh, desert, rain forest, and grassland. Along with a human **habitat** and an agricultural center, these biomes were meant to provide all the resources that Earth provides. **Sensors** were used to detect any problems. About 1,500 sensors in Biosphere 2 checked information like temperature and airflow. If each sensor took 360 readings per hour, how many readings did all sensors record in 1 day? To solve this word problem, multiply the number of sensors by the number of readings per hour. Then, multiply this number by the number of hours in a day.

```
     1,500  sensors
   x   360  readings per hour
     0 000
    90 00
 + 450 0
   540,000  readings per hour


   540,000  readings per hour
 x      24  hours a day
 2 160 000
+ 10 800 00
12,960,000  readings per day
```

The sensors recorded 12,960,000 readings per day!

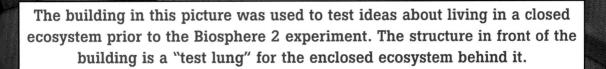

The building in this picture was used to test ideas about living in a closed ecosystem prior to the Biosphere 2 experiment. The structure in front of the building is a "test lung" for the enclosed ecosystem behind it.

Within the Biosphere

Human Habitat

In 1991, 4 men and 4 women from the United States, England, Germany, and Belgium were selected to live inside Biosphere 2 for 2 years. These 8 people needed a special space to live while they conducted research—a human habitat. Their living quarters included apartments, a kitchen, a dining room, a computer area, and even a library. Once this crew entered Biosphere 2 on September 26, 1991, they did not expect to return to the outside world for 2 years. They were only allowed to use resources within Biosphere 2. They could communicate with outsiders using telephones and video cameras, or by simply looking through a glass barrier.

Only 1 of the 8-member Biosphere 2 team left during the 2 years. She was absent only 5 hours after she injured her hand and needed additional medical treatment outside Biosphere 2.

The crew was given a 3-month supply of food at the beginning of their stay. This was the estimated amount of time needed to start growing food for themselves. The crew was put on a special low-calorie, high-**nutrient** diet to see if eating certain foods could promote a longer, healthier life. For the first 10 months, each member was allowed 2,000 calories a day. However, at the end of this time, the crew was hungry and tired, and had lost weight, so they were allowed a 10% daily increase in calories. How many calories a day did each crew member consume after the increase?

To solve this problem, change the percent to a decimal by placing the number over 100 and dividing.

$$10\% = \frac{10}{100} = \quad 100\overline{)10.0} \atop \underline{-10\ 0} \atop 0 \quad\ \ ^{.1} = .1$$

Multiply the original number of calories by the decimal number.

	Add the extra number	
2,000 calories	of calories to the	2,000 calories
x .1	original number of	+ 200 calories
200.0 additional calories	calories consumed by	2,200 calories
	each crew member.	

Each crew member was allowed a total of 2,200 calories a day.

Ocean

The ocean biome was also represented in Biosphere 2. Because oceans cover most of Earth, studying these waters and their inhabitants offered a chance to discover how to improve the overall health of the planet. The ocean of Biosphere 2 consisted of hundreds of thousands of gallons of water in a tank that was 25 feet deep. It took 38 trucks to transport 100,000 gallons of water from the Pacific Ocean. The rest of the water was a mixture created by scientists. Many kinds of fish and mammals lived here, along with ocean organisms like algae. Among other things studied in the ocean biome was a coral reef. Coral reefs, which house communities of organisms in Earth's oceans, are swiftly disappearing, along with many species that live in them. Biosphere 2's scientists **monitored** the growth of their coral colonies. After 2 years, the amount of coral colonies increased.

Let's say the ocean biome initially had 767 hard coral colonies and 131 soft coral colonies. If, after 2 years, 87 new coral colonies were found, how many coral colonies were there altogether? The word "altogether" appears again in this problem. Do you remember what to do? Even though the problem involves hard, soft, and new coral colonies, the problem asks for the total number of coral colonies. To solve this problem, add all of the numbers together to find the total number of coral colonies.

```
  767   hard coral colonies
  131   soft coral colonies
+  87   new coral colonies
  985
```

There were 985 coral colonies altogether.

Biosphere 2's ocean biome had a small beach where the crew could relax.

Marsh

Another biome included in Biosphere 2 was a marsh. A marsh is a wetland with grassy vegetation. It is usually located between a body of water and dry land. A marsh can contain freshwater or salt water. Biosphere 2's marsh was called an **estuary** because both freshwater and salt water flowed into it, making it a good habitat for many kinds of fish, snails, crabs, oysters, and other sea creatures. Among the most successful plants in the marsh were mangroves, which are trees and shrubs with many thick roots that grow above the water and provide food for many water organisms.

More than 250 species were introduced into the marsh, mostly taken from Everglades National Park in Florida. Let's say that fish represented 20% of the total species in the marsh biome and mangroves represented 16% of the total species. What number of species did these organisms represent altogether? First, find the numbers of species that fish and mangroves represented separately, then add them together. For each problem, we need to change the percent into a decimal by dividing the number by 100. Then, multiply the total number of species (250) by the decimal. Add the numbers of species of fish and mangroves together to find the total number of species they represent.

$$20\% = \frac{20}{100} = \quad 100\overline{)20.0} \quad \xrightarrow{\;.2\;}$$

$$\begin{array}{r} -20\,0 \\ \hline 0 \end{array}$$

250 total species in marsh
$$\begin{array}{r} \times\ .2 \\ \hline 50.0 \text{ species of fish} \end{array}$$

$$16\% = \frac{16}{100} = \quad 100\overline{)16.00} \quad \xrightarrow{\;.16\;}$$

$$\begin{array}{r} -100 \\ \hline 600 \\ -600 \\ \hline 0 \end{array}$$

250 total species in marsh
$$\begin{array}{r} \times\ .16 \\ \hline 1500 \\ +250 \\ \hline 40.00 \text{ species of mangroves} \end{array}$$

$$\begin{array}{r} 50 \text{ species of fish} \\ +\ 40 \text{ species of mangroves} \\ \hline 90 \text{ species} \end{array}$$

Fish and mangroves would represent 90 species out of a total of 250 species in the marsh. Can you think of another way to solve this problem?

Desert

You may picture a desert as mostly sand, rock, and scattered cactuses. A true desert is any dry land that receives less than 10 inches of rainfall each year. Biosphere 2's desert biome was modeled after a coastal desert. This means that it is not always as dry as other deserts and has shrubs and grasses. Because the rain forest and ocean biomes were not far away, Biosphere 2 had difficulty supporting a truly dry desert. The humidity and moisture from the ocean and rain forest biomes entered the desert biome in the form of fog.

You may also think that a desert is always hot. Most desert climates, including Biosphere 2's desert biome, have both hot and cold temperatures. A desert biome can be 100°F on a summer day and 35°F on a winter night. Let's say Biosphere 2's desert sensors recorded the temperature at 12:00 P.M. for 6 days in a row during the summer months. The sensors measured the following temperatures: 95°F, 90°F, 92°F, 91°F, 95°F, 92°F. What was the average temperature? To figure out an average, you will need to add all temperatures together and divide by the total number of temperatures (6).

$$
\begin{array}{r}
95°F \\
90°F \\
92°F \\
91°F \\
95°F \\
+\ 92°F \\
\hline
555°F
\end{array}
\qquad
\begin{array}{r}
92.5 \\
6\overline{)555.0} \\
-\ 54 \\
\hline
15 \\
-\ 12 \\
\hline
30 \\
-\ 30 \\
\hline
0
\end{array}
$$

The average temperature would be 92.5°F.

Some desert areas of Earth have grown larger and some have become smaller over time. The Biosphere 2 team studied their desert biome closely to understand more about why desert environments change.

Rain Forest

A rain forest biome is a wooded area with an average rainfall of 100 inches per year. A rain forest can be **temperate** or **tropical**. Biosphere 2's rain forest had high temperatures, like the tropical rain forests near the equator. Rain forests are important to study because they are home to almost half of all plant and animal species in the world! Rain forests are quickly disappearing due to human actions, and many of the species in them are in danger of becoming extinct.

What is the difference between the average annual rainfall of a rain forest and the greatest amount of annual rainfall a desert usually has? The word "difference" is usually a clue that we will need to subtract. To figure out this problem, you need to look at page 16 to see that a desert gets no more than 10 inches of rainfall each year. Therefore, we will subtract this number (10 inches) from the average amount of rainfall in a rain forest (100 inches).

> 100 inches of rainfall in rain forest
> − 10 inches of rainfall in desert
> 90 inches
>
> The difference in rainfall is 90 inches.
> What is the difference in feet?

Biosphere 2's rain forest had a 50-foot "mountain."
This structure was made of concrete and plaster and had
pockets of soil in which plants could grow.

Grassland

Another biome in Biosphere 2 was a grassland. Grasslands feature small trees and shrubs located near water. Grasslands are found on every continent except Antarctica. In Biosphere 2, over 40 kinds of grass grew on a small area overlooking the ocean biome. Species of grass from Africa, South America, and Australia were brought to Biosphere 2. Several plants reproduced very well after 2 years. In fact, some plants—such as the African acacia trees—grew so well that they blocked some sunlight from coming through the glass. Therefore, other plants like fruit and nut trees could not grow as well.

Let's say that for every African acacia tree that grew, 3 fruit and nut trees did not survive. We can show this relationship with the ratio 1:3, or $\frac{1}{3}$. If 11 African acacias grew over 2 years, how many fruit and nut trees did not survive? To figure out this problem, we need to set up a **proportion**. A proportion is a way of comparing 2 equal ratios. If we compare 11 African acacia trees to an unknown number of fruit and nut trees, with **t** being the unknown, the problem would look like this:

$$\frac{1 \text{ acacia tree survived}}{3 \text{ fruit/nut trees did not survive}} = \frac{11 \text{ acacia trees survived}}{t \text{ fruit/nut trees did not survive}}$$

To solve the word problem, we need to perform an operation called cross-multiplying.	$\dfrac{1}{3} \diagdown \diagup \dfrac{11}{t}$ $1 \times t = 11 \times 3$	$t = 11 \times 3 = 33$ For every 11 acacia trees that survived, 33 fruit and nut trees did not survive.

Since the vines and tall trees were killing the fruit
and nut trees, the scientists needed to control the plant population to
keep this part of their food supply from becoming "extinct."

Agricultural Biome

    The last section in Biosphere 2 was a farming area called the agricultural biome where crew members produced about 80% of their food. The crew grew food without chemical **pesticides** to combat pests or artificial fertilizers to make the soil more fertile. Instead, they introduced insects, like ladybugs, that fed on certain crop pests. They also rotated the position of crops so they could use nutrients from different plots of soil. They had $\frac{1}{2}$ acre on which to grow crops like potatoes, peanuts, rice, grapes, and bananas. How many square feet are in $\frac{1}{2}$ acre? We learned earlier in this book that 1 acre equals 43,560 square feet. One-half acre is $\frac{1}{2}$ this amount. Divide 43,560 by 2.

$$
\begin{array}{r}
21{,}780 \\
2{\overline{\smash{)}\,43{,}560}} \\
-4\phantom{3{,}560} \\
\hline
03\phantom{{,}560} \\
-2\phantom{3{,}560} \\
\hline
15\phantom{{,}60} \\
-14\phantom{{,}60} \\
\hline
16 \\
-16 \\
\hline
00
\end{array}
$$

One-half acre is 21,780 feet.

Agriculture accounted for about $\frac{1}{3}$ of the workload for each crew member every day, leaving only $\frac{2}{3}$ of their workday for all 6 other biomes! If the workday was 9 hours long, how much time was given to the agricultural area?

In the first year alone, the agricultural biome produced 7 tons of food for the crew and 15 tons of food for the animals, which included goats, pigs, and chickens. Over 80 different kinds of crops grew despite the fact that Arizona was unusually cloudy during the years of the experiment. In fact, the cloudy conditions caused the scientists to experiment with new crops that could grow in the shade.

One ton is equal to 2,000 pounds. How many pounds of food did the crew raise altogether? This is a 2-part word problem. First, we need to convert tons to pounds. Then we need to add the amount of pounds produced for the crew and the amount of pounds produced for the animals.

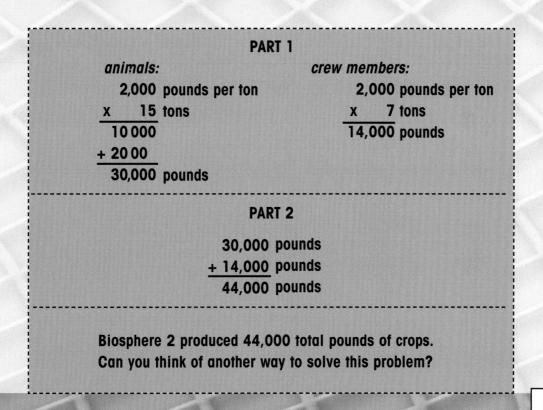

PART 1

animals:

2,000 pounds per ton

x   15 tons
———————
10 000
+ 20 00
———————
30,000 pounds

crew members:

2,000 pounds per ton

x   7 tons
———————
14,000 pounds

PART 2

30,000 pounds
+ 14,000 pounds
———————
44,000 pounds

Biosphere 2 produced 44,000 total pounds of crops.
Can you think of another way to solve this problem?

Success or Failure?

When the crew stepped out of Biosphere 2 on September 26, 1993, many people debated whether the 2-year experiment was a success or failure. One reason some thought it failed was that the oxygen levels had dipped so low at one point that the crew felt tired, slept badly, and breathed poorly. At the beginning of the experiment, the oxygen level was 21%; the oxygen level dropped as low as 14.5%. Another decrease in oxygen could have been deadly. In January of 1993, 15.7 tons of oxygen had to be pumped into the building, even though Biosphere 2 was supposed to operate without outside assistance. This brought the oxygen level back up to 19%. Scientists studied the effect of the increased oxygen level on the crew. Scientists discovered that air containing between 16% and 19% oxygen allowed crew members to function normally. This information is helpful to people like mountain climbers and deep-sea divers, who need to perform well in low-oxygen environments.

The average person breathes about 550 liters of oxygen per day. At this rate, how many liters of oxygen does a person breathe over a year? To figure out this problem, multiply the number of liters of oxygen breathed per day (550) by the number of days in a year (365).

$$
\begin{array}{r}
550 \text{ liters per day} \\
\times\ 365 \text{ days} \\
\hline
2\,750 \\
33\,00 \\
+\ 165\,0 \\
\hline
200{,}750 \text{ liters per year}
\end{array}
$$

The average person breathes about 200,750 liters of oxygen a year. About how many liters of oxygen do 8 people breathe during 2 years? To find the answer, multiply 200,750 liters by 8 people. Then multiply your answer by 2 years.

One possible reason for the low oxygen levels in Biosphere 2 is that tiny organisms living in the soil were using up more oxygen than originally planned while breaking down dead organic materials.

The Biosphere 2 experiment was set up in part to observe how biomes affect each other. Air flowed freely within Biosphere 2, which could have distorted the results of experiments and affected the growth of organisms. On Earth, these biomes do not exist so close to each other. Some critics of Biosphere 2 thought **partitions** should have been placed between the biomes. Also, the glass structure, no matter how transparent, still reduced the amount of natural light. This would favor some plants over others that might grow better under direct sunlight.

Although some species like honeybees did not survive in Biosphere 2, others like ants, crickets, and ladybugs increased in number.

In addition, only certain organisms were introduced into each biome. No animal or insect too large, too poisonous, or too difficult to feed was considered. Therefore, the survival rate of species could not be applied to the outside world. During the 2-year period, a percentage of organisms was expected to "become extinct." Let's say that 15 of the 45 species did not survive. What percent of 45 is 15? First, we need to set up a proportion. Then we need to cross-multiply those ratios.

$$\frac{t}{100} = \frac{15}{45}$$

$45 \times t = 100 \times 15$

$(45 \times t) \div 45 = 1{,}500 \div 45$

Now, divide 1,500 by 45.

```
        33.33
45) 1,500.00
    −135
     150
    −135
     150
    −135
     150
    −135
      15
```

This answer is called a repeating decimal because the number 3 will repeat without end. Let's round the answer to the nearest hundredths place.

Fifteen is 33.33% of 45. Of the total 45 species of insects, 33.33% became extinct.

What fraction of the 45 species of insects became extinct?

Before the second crew entered, 34 species were brought into Biosphere 2's ocean biome, shown above. Ten of these species were new additions to the Biosphere 2 ocean. What fraction of species was new to the ocean biome? See the box on page 29 for the answer.

The Biosphere 2 scientists made many attempts to fix problems before the second crew was sent in. In 1994, however, this second mission had a shortened stay in the building. It was decided that the closed ecological system approach had too many problems. Never again have people lived within a closed environment like Biosphere 2, though it has remained a valuable research laboratory for scientists.

Although Biosphere 2 had its critics, it is hard to deny that it was a breakthrough experiment. Never before had people survived for so long in a closed ecological system. The problems that occurred inspired research in many subjects. One scientist stated that it is impossible to have a perfect experiment, especially one that tries to include so many aspects of Earth's life cycles. Too many **variables** exist in nature that we do not yet understand. The Biosphere 2 experiment warns us to be careful of our actions on this planet because we cannot create another one.

To solve this problem, place the number of new ocean species above the total number of species added to the biome.

$\dfrac{10}{34}$ 10 new ocean species / 34 total ocean species

Simplify the fraction by dividing both the top number and the bottom number by 2.

$$\begin{array}{r} 5 \\ 2\overline{)10} \\ -10 \\ \hline 0 \end{array} \qquad \begin{array}{r} 17 \\ 2\overline{)34} \\ -2 \\ \hline 14 \\ -14 \\ \hline 0 \end{array}$$

$\dfrac{10}{34} = \dfrac{5}{17}$

Of the 34 total species brought into the ocean biome, $\frac{5}{17}$ were new to Biosphere 2.

Biosphere 2 Today

After the second mission ended, Columbia University took control of Biosphere 2 and managed it for several years. Students who wished to study the environment using a unique laboratory and the latest technology were welcomed to use the Biosphere 2 facilities. Although it was decided that people would no longer live in Biosphere 2, its different biomes were still useful research centers for many professionals who needed to conduct experiments in a controlled setting.

Columbia University has decided to end its programs at Biosphere 2, and the center's future is uncertain. With new developments and discoveries in space, perhaps someday the knowledge gained from the Biosphere 2 experiment will help to build bases on the moon or other planets. To accomplish this, we must be able to identify and solve problems that arise in life, like the word problems we solved in this book. For now, hundreds of thousands of visitors tour Biosphere 2 each year to learn more about the science of Biosphere 1—Earth.

Glossary

biome (BY-ohm) A major type of ecological community, such as a rain forest or a desert.

environment (in-VY-ruhn-muhnt) The living things and conditions that surround an area.

estuary (ESS-chuh-wear-ee) A water passage where a saltwater tide meets a freshwater current.

habitat (HAA-buh-tat) A place to live.

monitor (MAH-nuh-tuhr) To watch for a special purpose.

nutrient (NOO-tree-uhnt) A substance that promotes growth.

organism (OHR-guh-nih-zuhm) A living thing.

partition (pahr-TIH-shun) Something that divides something else into sections.

pesticide (PES-tuh-syd) A substance used to destroy pests.

proportion (pruh-POR-shuhn) A mathematical statement that shows 2 ratios to be equal.

sensor (SEHN-suhr) A device that measures characteristics of an environment and sends information to a receiver.

temperate (TEHM-puh-ruht) A climate that is not too hot or too cold.

tropical (TRAH-pih-kuhl) Having a moist climate that supports year-round plant growth.

variable (VAIR-ee-uh-buhl) Something that can change.

Index